695

53 Interesting Ways To

Teach Mathematics

Interesting Ways To Teach

In the same series:

53 Interesting Ways To

Teach Mathematics

Ruth Hubbard

Technical and Educational Services Ltd.

First published in 1991 by
Technical and Educational Services Ltd.
37 Ravenswood Road
Bristol BS6 6BW
UK

ISBN 0 947885 60 9

Printed by Billings & Sons Ltd., Worcester

About Ruth Hubbard

Ruth lectures in the School of Mathematics at the Queensland University of Technology in Brisbane, Australia. Most of her ideas for facilitating learning have come from her observations of students struggling to learn mathematics. For a number of years she managed a Remedial Mathematics Facility and produced remedial materials to try to help those who were having the greatest difficulty. After much thinking and planning, she abandoned lectures as a means of imparting information in some of her courses. For a number of years her students have been working, usually in groups, from texts and study guides. She enjoys, and gets great satisfaction from, inventing new ways of responding to students' needs.

Contents   Page

Preface

Along with the other books in this series this book aims to help people who teach in tertiary institutions. Most such people do not have any formal training in education so theories of learning have been translated into practical class activities. The book may also be of use to staff developers, most of whom do not have a mathematical background.

Why single out mathematics teaching? Mathematics is a very old discipline so the teaching of mathematics has a long tradition. This probably explains why our teaching is so conservative and very slow to respond to changing circumstances. And mathematics itself is different from other disciplines. It deals with abstraction, it has its own symbolic language and it is perfect in the sense that new theories do not change it.

Mathematics is studied by a large proportion of tertiary students. Mathematics is perceived to be difficult and the prerogative of clever people. And since students start to learn mathematics when they are very young they come to universities with a large body of mathematic experience, both good and bad. They have developed attitudes to mathematics and methods of operation which are difficult to change.

Lecturers in mathematics now face a situation in which their students may have been taught in schools in a fairly traditional way, working from established text books under tight control by a teacher. However, progressive methods are now also common in schools, and students may have be used to 'discovery' methods involving group problem-solving and a lot of talk. Indeed, learning through talk is strongly advocated in the *National Curriculum* currently being introduced into the UK education system.

Lecturers in further and higher education, whether working in the UK or in other countries in which students share similar learning experiences, should therefore be alert to the possibility that formal lectures are as strange to some of their students as collaborative group work may be to others.

For these reasons I have urged caution throughout this book. Whenever you want students to do something new in a mathematics class you have to convince them that the change is worthwhile and to show them how to do the new task. Of course you have to be convinced too and prepared to persevere in the face of opposition from some students. It is not a good idea to introduce too many innovations at once.

Some of the activities in this book are mathematical adaptations of ideas from other books in the TES *Interesting Ways To Teach* * series and from an earlier book, *Teaching Students To Learn* (Gibbs, 1982). Many other activities arose spontaneously in the course of my teaching. They were then discussed with some sympathetic colleagues and tried again more formally. Of course many of the underlying ideas originate in the work of a wide variety of mathematicians and educators. Whenever I was aware of the inspiration for an activity I have acknowledged it and I apologise for any omissions. My contribution has been to devise class activities which work and which can make mathematics teaching and learning more effective, satisfying and pleasurable.

Many colleagues who teach mathematics have read through early drafts of the book, suggested amendments and offered sound practical advice and peer support. To them all I offer my sincere thanks. Finally, I would like to acknowledge the inspiration, encouragement, sensitive advice and helpful support I have received from John Davidson (Oracy Video), Graham Gibbs and Trevor Habeshaw in the final production of this book.

Ruth Hubbard
August 1990

Reference

Gibbs, G.   *Teaching Students To Learn* Open
University Press. 1982.

* Details of how to order TES publications can be found at the end of this book.

12

Designing courses

1   Keller plan courses

2   Small group courses

3   Problem-based courses

4   A mixed model

5   Remedial courses

Designing courses

Lecture-based courses with some kind of tutorial support still dominate mathematical instruction. However, in response to problems such as a more heterogeneous student body and advances in technology, other models have been proposed. None of the new designs have so far posed a serious threat to the lecture-based model, but they have enthusiastic adherents and are worthy of consideration.

The rapidly increasing number of students undertaking tertiary study, including some mathematics or at least statistics, has produced a need for remedial, and catch-up courses. Since these are comparatively new phenomena with no established tradition, many innovative ideas have been used in their design.

Some of the most widely used alternatives to lecture-based courses will be described in this section as well as some remedial programmes.

Keller Plan Courses 1

Many variants of Keller's original proposal have been used in mathematics, both for mainstream courses and for remedial ones. The essential features are that to a large extent students work at their own pace and that they present themselves for mastery tests when they think they are adequately prepared. They must demonstrate their mastery of a segment of the course before they can progress to the next segment. Student proctors or other tutors mark the tests individually in the presence of the student and provide some instruction during the marking process. In the main students learn from study guides. Frequently there are no lectures.

Courses designed along these lines clearly have some advantages over the standard lecture course. The main advantage is that very capable students can progress through the course very quickly while slower students take their time. The mastery tests are supposed to ensure that students learn the material thoroughly. The individual marking allows misconceptions and weaknesses to be identified and corrected immediately. Even with modified Keller plan courses a number of problems have been identified and you will need to address these when planning a course.

a   Some students tend to procrastinate so it may be necessary to set deadlines by which certain segments of work must be completed. Otherwise some penalties must be imposed.

b   There is a lot of work involved in the initial preparation of study guides and tests. Because of the mastery requirement some students need to repeat tests, so many versions of each test are required. However, once this work is done the material is available for use in later years.

c   Security can be a problem when the tests are administered to large numbers of students. Computer-generated tests from a test-bank can overcome this problem.

d   One-to-one test marking and tutorial assistance require a considerable investment in tutor hours.

e   Some system of training and guidance for student proctors is usually necessary. The question of payment of proctors, or some other credit arrangement, may not prove unsurmountable if the proctor can be convinced that both the recipient and the donor will benefit from this activity.

f   An efficient system of record-keeping, probably computer-based, is required.

g   Students who repeatedly fail mastery tests need to be identified and given additional help.

h   The system encourages students to learn the material in small pieces. A final examination may be needed to integrate what has been learned.

i   All learning is on an individual or one-to-one basis. There is no opportunity for discussion. (This is a common problem with non-Keller programs as well.)

More details on these problems and ways to overcome them can be found in the second reference below.

References

Keller, F.S.   Goodbye teacher
                      Journal of Applied Behavioral Analysis 1, 79, 1968.

Brook, R.J.   The Evolution of a Keller Plan Service Statistics Course
& Thomson, P.G.   *Programmed Learning and Educational Technology,*
                      Vol. 19, No. 2, 1982.

Small group courses 2

Weissglass gives the following simple definition of learning: *Learning consists of evaluating new information in relation to information that's already understood and storing it in a form that's available for use in new situations.*

If you are prepared to accept this definition then you will want to choose a course structure which facilitates this view of learning. Small group work is one way to guide students towards this kind of learning. You will need to give some thought to how these groups are established, timetabled and supervised, and students will need to be clear about the working arrangements, e.g. whether they are to be compulsory or self-paced.

Most people who have tried this approach recommend that the class be divided into groups of 3 or 4 students. The groups are provided with worksheets which are both a source of factual information and of questions for group investigations. Clearly the worksheets need to be very thoughtfully prepared so that students working through them actually learn in the desired manner. Some courses of this kind have been based on the Open University materials, which have essentially been produced to promote individual learning. However, group learning has many advantages over individual study. Some of these advantages and also advantages over lecture courses are outlined below, as well as in Item 20: *Group tutorials.*

a   Groups can work through the worksheet at their own pace. This is impossible in a lecture.

b   Students learn effectively from their peers. There is never sufficient time for everyone to engage in one-to-one learning with the instructor, but small groups make it possible for students to engage in one-to-one discussion and problem-solving with their peers.

c   The instructor can concentrate on facilitating learning within the groups rather than on delivering factual material.

d   Students have the opportunity to read, write and talk mathematics. In particular, when one student explains something to another she deepens her own understanding.

e   It is less threatening to ask questions of one's peers in a small group than to ask the instructor in front of the whole class.

f   By trying different approaches or explaining their point of view to others students develop an understanding of how mathematics really works.

g   It is easier to learn in a friendly, non-threatening environment. In lectures, and when studying alone, students often feel isolated and insecure.

h   Doing mathematics can become an enjoyable experience instead of an obstacle to be overcome.

In his article Weissglass provides a useful handout for students embarking on a small group course for the first time. In a similar article Barbut, who has based his small group teaching on the model proposed by Finkel and Monk, offers examples of worksheets in abstract algebra.

References

Barbut, E.   Abstract algebra for high school teachers: an experiment with learning groups, *Int. J. Math. Educ.Sci Tech.*, Vol. 18, No. 2, 1987.

Finkel, D.L. &   *Teachers and learning groups: Dissolution of the Atlas*
Monk, G.S.   *Complex.* In *Learning in Groups. New Directions in Teaching and Learning*, No. 14, C. Bouton and R.Y Garth eds. Jossey-Bass, 1983.

Weissglass, J.,   Small Groups: An Alternative to the Lecture Method *Two Year College Mathematics Journal*, Vol. 7, 1976.

Problem-based courses   3

In a non-mathematical context a problem-based course implies that the problem is encountered first in the learning sequence before any preparation or study has occurred. On the other hand most mathematicians argue that the knowledge and skills must come first and furthermore that only knowledge and skills with which the student is already very familiar are available for problem solving.

Problem-solving can be used in a variety of ways, for example,
-    to enable students to practise already taught knowledge and skills;
-    to preface, or give urgency to the learning of knowledge and skills;
-    to put all learning into 'real' contexts;
-    to give relevance and coherence to learning.

As a result problem-based courses tend to be optional extras rather than genuine alternatives to mainstream courses. This attitude to problem based courses may change now that mathematical investigations are becoming an integral part of school mathematics. There are three basic kinds of problem-based courses which you could use as models from which to design your own course.

a   Mathematical modelling courses which deal with 'real world' problems. Students may have to collect data as well as constructing a model, testing it and so forth.

b   Courses in solving abstract mathematical problems.

c   Courses which combine (a) and/or (b) with computer programming to produce numerical solutions to problems.

If you wish to pursue any of these alternatives you will need two things. The first is a supply of good problems. Fortunately there are plenty of books of good problems of all levels of difficulty for the kinds of courses described above. One reference for each kind is given at the end of this item. Secondly, you will have to design the course so that the students rather than the instructor, solve the problems. Several people who have experimented with such courses claim that the most difficult part of the process for them is to stand back and keep quiet.

You will also have to make a decision about how much or how little instruction on problem solving you give, for which help can be found in Item 28: *Problem Solving*, and about how the students are to work, individually or in groups. Group work has generally been found most effective, but this raises problems with assessment, for example, how to assess the contribution of an individual to a group solution. A simple answer is to assess the work on a pass/fail basis, without awarding grades. This leads to the question of examinations for problem-based courses. Since a course without a final examination appears to lack status, examinations have been devised. However it is unlikely that many students could individually solve a substantial problem under examination conditions.

There are also decisions to be made as to whether solutions are to be presented orally or as written assignments, whether to use a few large problems or many small ones and so on.

If you decide to design a problem-based course you will clearly have a number of problems to solve yourself!

References

| Beilby, M. & McCauley, G. | *Introduction to Computational Mathematics,* Scottish Academic Press, 1986. |

Berry, J.S. et. al.   *Mathematical Modelling Courses*, John Wiley, 1987.

Mason, J., Burton, L. & Stacey, K.   *Thinking Mathematically*, Addison Wesley, 1985.

A mixed model 4

A mathematics course usually has more than a single objective. For example you may hope that by taking your course students will develop their manipulative skills, some facility at modelling problems and confidence in applying mathematics in other subjects. It is difficult to address these objectives and others that you may add using only one method of instruction. Keller plan courses, for example, are not very suitable for teaching problem-solving, improving motivation or developing confidence in mathematics.

No course design will suit all your students. Some may prefer to study individually, some want to be fed a continuous diet of lectures, while others enjoy working in groups. In addition, many students have a very narrow view of mathematics. By presenting it in a variety of modes their view may be broadened.

It is not possible to satisfy all these demands all of the time, but you might give some thought to a course in which different approaches are used to meet the varied objectives. Manipulative skills require practice, something a student can do on her own, so this objective could be achieved by individualised learning and periodic tests. Some of the factual material could also be given to students to read on their own.

Lectures can be given at various times for specific purposes such as to provide the motivation for a new topic, to give an overview or a review and to communicate a more personal approach to the material. Tutorial sessions can be used to help weaker students to keep up with the individual work. Problem solving and investigative work are best done in groups using carefully prepared materials. Oral skills, written communication skills and self-confidence can also be developed as part of group work.

If you would like to design a course which involves several of the approaches suggested here you will need to do a lot of planning and preparation. There is so much work involved in establishing such a course that it usually needs to be a co-operative venture with several people pooling their expertise. The bonus is that once everything is prepared it is available for future years and may only require minor modifications. You will have to think very carefully about how to achieve the correct balance between the different aspects of the course. The timing of lectures, tests, and problem sessions is also important. Students may have to complete certain sections before they can appreciate lectures and take part in problem solving sessions.

If the whole plan becomes too complicated, inexperienced students may find all the different demands on them rather bewildering. Some counselling may be required. Although mixed model courses appear to have many advantages, there are not many examples in the literature to help you. A rather brief report on one such course is given in the reference below.

Reference

Pickering, R.A. &   An investigation into the teaching of mathematics to
Watson, F.R.,   undergraduate economics students
                        Int. J. Math. Educ. Sci. Technol., Vol. 17, No. 3, 1986.

Remedial courses 5

Over the past 10 to 15 years remedial programs in mathematics have been developed in a large number of tertiary institutions. Because these courses are new they are in a sense free from traditional teaching methods, so many different models have emerged. Most remedial courses were started by dedicated people who were keen to do something to help the ever increasing numbers of students coming to their classes without adequate preparation. Some of the programs were originally designed for special groups, mature aged students, women or other disadvantaged groups in society. However, as time has gone by, a larger proportion of the regular intake of students seems to require remediation.

Some of the major types of courses are described below, but there are many local variations and combinations of course types in existence.

a   Lecture courses in which the instructor is the main provider of information, often delivering it in much the same way as it was delivered at school. It is even more difficult to find a pace that suits all students in a remedial course than in a standard course. Some students, particularly more mature ones, learn very quickly once the material is presented to them in a systematic way. Other students require a lot of time to feel comfortable with new concepts. Also lecture courses do little to help students whose problems stem from 'maths anxiety': in fact, they may reinforce these feelings.

b   Independent learning courses which provide students with printed materials, tapes, videos and computer packages from which the student is supposed to learn alone. This is the cheapest way to run a remedial program but may not be the most effective. Unless there is some mandatory testing or counselling only highly motivated students are likely to benefit from this approach. If testing is part of the program then the course becomes some variant of a Keller Plan. A useful self-help book for maths-anxious students is mentioned in the references.

c   Group work courses have been designed specifically to deal with 'maths anxiety' and to improve students' confidence both as mathematicians and as people. Students are provided with materials as in (b) but there are also structured group activities. These consist of investigations which can lead to a deeper understanding of mathematical ideas. By working in groups students are able to give each other support and encouragement. Burton has described a course of this kind for disadvantaged groups. Of course a program of this kind is much more difficult to plan and more expensive to run than (a) and (b) above but it has the capacity to offer greater rewards.

d   No course at all, but the establishment of a mathematics learning centre which can be used by students on an individual basis. The centre contains learning materials and is staffed by tutors who can be consulted on mathematical problems of all kinds. Attendance is usually voluntary. Personal contact with tutors, particularly effective in motivating students to undertake remedial work, supplements available study materials.

Of course it is possible to offer remedial mathematics in several modes so that students can choose the learning method they prefer. Whatever model you use it is important that the instructors involved are sympathetic to students' difficulties and are able to make the students feel that they can make a fresh start and achieve some success in mathematics.

References

Burton, L.	From failure to success: Changing the experience of adult learners of mathematics.
	Educational Studies in Mathematics 18, 1987.
Patterson, D. & Sallee, T.,	Successful remedial mathematics programs: Why they work.
	The American Mathematical Monthly, 93 (9), 1986.
Tobias, S.,	Overcoming math anxiety
	Houghton Mifflin, 1980.

Giving lectures

6   The beginning

7   The middle

8   The end

9   Lecture-tutorials

10 Buzz groups

11 Overview lecture

12 Learning from reading

13 Uncompleted handouts

14 User-friendly handouts

15 Misprints

Giving lectures

Lecturing is still the most common teaching method in mathematics despite the unequivocal evidence that students find it very hard to learn from them. At best, lectures can be exciting and inspiring, but usually they are not. At worst, lectures are an expensive and time-wasting reprographic technique. Nevertheless, you will probably continue to lecture, so here are some ways to make your lectures more effective.

The first three items deal with aspects of conventional lectures in which the lecturer explains what he or she is writing on the blackboard or presenting on prepared transparencies. The next group of items suggest ways of varying the standard approach by giving the students a more active role in the proceedings.

The three final items concern handouts produced to accompany lectures.

The beginning 6

You probably skim through your lecture notes just before the lecture begins to get yourself in the right frame of mind. Don't forget that your students haven't done this, unless you have them very well trained! You will need to organise their thoughts so that they can appreciate your lecture. Some of the following techniques should help:

a   Display an outline of your lecture on the overhead projector as students arrive

b   Explain how this lecture fits into the terms work, how it follows from previous lectures and where it is leading

c   If you are starting on a new topic, find an interesting example to motivate the students

d   Refer students to the relevant pages in the text or handout

e   Remind students of the meaning of any notation you will use and the meanings of words they may have forgotten

f   Educational theory suggests that genuine learning can only take place when new concepts can be linked with existing ones in the student's mind. That is essentially what the following suggestions are designed to assist.

Make sure that the students have any concepts, rules, theorems that you intend to make use of fresh in their minds. Try reading through your lecture from the student's point of view to pick these points up. It is worth spending time at the beginning of the lecture to ensure that most students are equipped to understand the new ideas you will present.

For example ,if you are going to present partial fractions it is helpful to remind students of:

- the form of added fractions

- the fact that if two polynomials are equal their corresponding coefficients are equal for all values of x

The middle 7

The format of maths lectures is often a sequence of two basic patterns:

either

      definition   theorem   proof   examples,

or

      definition   examples   theorem   proof.

In the first pattern a general result is followed by special cases and in the second special cases are used to lead to a general result. The same patterns appear in the textbooks.

In spite of this well-known and rigid framework students notes are often confused.

You can help by making it absolutely clear where one thing finishes and the next one starts, e.g.

That completes the proof, now here is an example of how it can be used.

Throughout your lecture try to explain why you make certain decisions and why you don't make others. These are the things that the student will not find in a textbook but which he has to learn about from you. For example:

The denominator is the square root of a sum of squares so I think straight away of

$$1 + sinh^2 \theta = cosh^2 \theta$$

and will try to make a substitution based on that.

There are other things you can do during the lecture which are not usually found in written materials.

a   Break up a long proof or example into distinct labelled steps. Suppose you are finding the area under a curve as the limit of the sum of the circumscribed rectangles. First you need to find the area of the i' th rectangle, then you have to find the sum of these areas, then you have to find the limit of this sum. Tell the students in advance that these are the three stages and then indicate the end of each stage.

b   If you omit algebraic steps, say and indicate what they are, e.g.

Now we take out a common factor and write the expression with a common denominator.

c   Make it clear what is explanation and what you expect students to show in their solutions. This is important if you derive some results from first principles such as $\frac{d}{dx} \sin^{-1} x$ indicate that students may in future make use of this result.

d   Explain whether something follows only from the line above or whether a new idea has been introduced. Does the new idea come from the statement of the problem or from elsewhere? For instance in the example in (a) you will need a summation formula. This comes from outside this problem.

The end 8

The most important thing about the end of a lecture is that it finishes at the correct time. You should be watching the clock all through, but it is crucial towards the end. If you go over time your students will become restless, start to pack up and the end of your lecture may be very ragged. By finishing promptly you also give students who want to speak to you individually the opportunity to do so.

Don't leave important announcements to the end of the lecture when your students' concentration may be waning. Make announcements at the beginning or, if you have problems with latecomers (perhaps another lecturer has gone over time!), at some suitable stopping point in the middle.

Ensure that the students

a have all the relevant handouts

b know what follow-up work they should do after the lecture

c know when the next test or assignment is due

d know exactly what the next test will cover.

Try to finish at a logical place rather than in the middle of something. If you have taken longer over a topic than you intended, ask the students to complete the work from a handout.

It would be excellent to be able to finish each lecture on some exciting high note, but this is often very difficult to achieve. You may have to be content to finish efficiently. Sometimes you may be able to pose a problem which you can leave the students to think about or refer to an interesting or unusual application of the topic. If you leave the students with a problem, make sure you return to it at the beginning of the next lecture. If you just forget about it, so will the students the next time around.

Lecture-tutorials 9

People are able to concentrate hard only for a limited time. While it is not true for everyone in the class, where some concentrate for longer and others for shorter periods of time, a limit of 20 minutes is usually quoted as typical for students in higher education. Concentration spans can also vary with the complexity of the task, so maths teachers may have a particular problem. Their difficulty is further compounded by the fact that not all students concentrate for the same 20 minutes!

Mathematics lectures are normally scheduled to run for one or two hours without a break. If the students' concentration wavers after the first 20 minutes, it is pertinent to ask what benefit they get from the rest of the time. They can of course continue to take notes from the blackboard, transparencies or dictation. The problem of actually following what the lecturer is saying is particularly serious in mathematics because of the highly sequential nature of the subject. Each new theorem depends on definitions and theorems which have just preceded it, as explained in Item 7: *The Middle* . Even before a student loses concentration, he or she may cease to follow the lecturer's arguments as a consequence of not understanding the full implication of something discussed earlier in the lecture.

One way simultaneously to alleviate the problems of short concentration span and long sequences of new ideas, is to give students exercises to work during the lecture period. Give short exercises at suitable points every 10 to 15 minutes or a longer exercise after 20 minutes. Walk around while the students are working. They will appreciate your interest in their efforts and you will find out how your lecture is being received. The break from listening and note-taking will enable students who are getting lost to check back in their notes and try to fit together all the information they have been given. Simple exercises which encourage these activities will enable the student to consolidate what you have covered up to that point and to be prepared for what follows. Also, the mental activity involved in working through an exercise is very different from passive listening, so to some extent the student's ability to concentrate may be restored.

Tell the students the answer to these exercises before proceeding with the lecture or get students to check their answer with a neighbour, as in Item 10: *Buzz groups.* If the exercise is a direct application of the subject many students will have done it correctly and this will encourage them to take an interest in the rest of your lecture. It is better not to work through the exercise in front of the class because if you do so regularly students will not bother to do it themselves. They will just wait for you to produce the solution and much of the benefit of the technique will be lost.

Of course, if you break up the lecture hour with exercises you will not be able to cover as much material as if you talked non-stop. However, you must remember that what you attempt to cover and what the students are able to learn are two very different things. Since there are severe limits to what a student can learn in an hour it may be better to provide some of the material in a handout to supplement the lecture.

Buzz groups   **10**

Buzz groups are another way of breaking up a lecture into manageable pieces and of involving students in an active rather than a passive way.

Pose a question and ask students to discuss it in pairs or threes. To encourage discussion you will have to ask questions that are interesting and challenging. Simple, direct questions are useful if you want students to work individually, checking back in their notes for information, but are not suitable for starting discussion. Difficult questions which require a thorough mastery of the subject will leave the students looking blank and embarrassed. Try asking the students to make a conjecture or to generalise from some examples you have shown them. Conjecturing and generalising are important mathematical activities of which students are often quite unaware. As a result the first time you ask them to do this they may be somewhat puzzled. It always takes a little time for students to accept new ways of thinking.

Another task for buzz groups is choosing a suitable approach to starting on a question. For example if you have been teaching methods of integration you could give the class an integral and ask the groups to investigate and choose a suitable technique.

Before you ask students to discuss something in buzz groups arrange a signal which will indicate the end of the discussion. This will allow you to bring the class back to order.

Ask some students what their conclusions are. If you have chosen a really good question a real argument may develop! Try to use some of the ideas suggested by students as you develop the next part of your lecture. It encourages students to take their group discussions seriously if they expect that you will make use of their ideas.

Overview lecture   11

Since students have difficulty in quickly absorbing the detailed content of a conventional lecture it can be helpful to provide a clear overview.

The purpose of an overview lecture derives from Ausubel's description of an 'advance organiser'. Ausubel explains how these organisers can be used to

a   give a broad outline of the material which will be learned in detail later

b   show how the new material relates to existing knowledge

c   point out in what ways the new material is similar to or different from related concepts in the student's cognitive structure.

There are obviously many different ways of constructing a useful overview lecture, but since they are rarely used it may be best to illustrate with an example.

Outline of overview lecture on one-way Analysis of Variance.

1   Explain what is meant by treatments and how the existence of differences between treatments is established by comparing the variance between samples with the variance within samples.

2   Relate to previously studied hypothesis tests and to F as a ratio of variances.

3   Compare with the case when there are only two samples. In both cases hypotheses are about mean effects.

Reference

Ausubel, D.P.   *The Psychology of Meaningful Verbal Learning*, Grune and Stratton, 1963.

Learning from reading 12

As explained in Item 9: *Lecture-tutorials*, conventional extended lectures suffer from the problems of students' limited concentration and the sequential nature of mathematics. Whatever pace you choose for the delivery of your material, some students will find it too slow and thus boring. Others will find it too fast and quickly become lost. Lectures may not be the best way for students to learn mathematics. But what are the alternatives?

Before students can do any mathematics they must have some resources to call on. The *every student must discover everything for herself* approach, proposed by Bruner, can be very time consuming. Hence it is generally assumed that the student must have some source of facts and examples to draw on. Computer software and video have been used for this purpose to some extent but not with a great deal of success. The programmes require a considerable commitment both in personal and in resources terms, and in the end they may be no more helpful than a good book while being, as yet, a good deal less convenient. This leaves textbooks and handouts as the primary sources of information. But this introduces a new problem. Many students have great difficulty reading mathematical text and for good reason. Even when it is deliberately user-friendly (and usually it isn't) it is much more difficult to read than normal text, and students may need help, as in Item 37: *Reading*.

If you require students to read mathematics as a supplement to lectures or instead of lectures you may have to help them to develop some reading skills.

Providing all the information students need in the form of handouts is almost equivalent to writing a text. A less onerous task is to use a text, or several texts, to broaden students' experiences, and to write accompanying study guides.

The study guide can be used

    a   to tell students exactly which sections they should read;

b   to expand on difficult areas;

c   to emphasise what is important and what is ancillary;

d   to indicate which exercises should be attempted;

e   to provide solutions or answers.

All this sounds very like a distance education course. However, you have the advantage of regular tutorials during which two-way communication can take place. This may be more valuable than the one-way communication in a standard lecture.

Reference

Bruner, J.S.   The Act of Discovery
                     Harvard Educational Review 1961, 31.

Uncompleted handouts 13

One disadvantage of giving students a handout which contains the complete lecture is that the students may think they do not need to attend or, if they do attend, do not need to pay such close attention. A second disadvantage is that the notes in the handout are not their own. Even someone else's ideas in one's own handwriting seem more personal.

Uncompleted handouts can overcome both these problems, and at the same time serve as advance organisers. The student must pay attention in order to fill the blanks and the filled blanks are her own work. It is best to provide definitions and the statements of theorems in the handout because it is vital that students have perfectly correct versions of them.

The most obvious things to leave blank are the solutions to examples. This way the student sees the solution evolving instead of seeing it as a finished product. However, there are many other possibilities.

You could leave blanks for:

a   intermediate steps in a proof or the whole proof

b   diagrams which are parts of theorems or examples, as in Item 45: *Diagrams in problem solving*

c   different forms of a diagram

d   conjectures that you will encourage students to make

e   exercises, questions or tasks which form the basis of active learning in lecture-tutorials or buzz groups.

f   anything else that might have an element of surprise in it which you will explore during the lecture.

Example

Definition   A random variable   X   is a real-valued function of the elements of a sample space.

Example.   Suppose we have the sample space of families with 3 children, we used earlier,

S = {GGG, GGB,   }

Define the function   X   as the number of girls in a family so that

X(GGB) =   X(BBB) =

Now we will make up a table of the values of   X   and their associated probabilities.

          X
       P(X = x)

User-friendly handouts 14

Have a good hard look at the notes, exercises and solutions that you hand out to students.

Are they completely legible?

There is no point in adding to students' confusion with illegible handwriting or poor reproduction. If your exercises are copies of copies of copies it is time to start again. Because of the concise nature of mathematical text students cannot guess what an illegible word or symbol should be.

Are they free of typographical errors?

Typographical errors occur most frequently because typists do not fully understand mathematical notation. If you write your script carefully the typist will be more likely to copy accurately. If your handwriting is legible, typing is not necessary. Handwritten notes look friendlier. If your handwriting is awful, perhaps you could learn to use a mathematical word-processing package.

Ask a colleague to proof-read your handouts. He or she will read what is actually there, not what you think is there.

Remember that students can waste hours trying to follow an argument which involves an error.

There are, however, ways of using typographical errors creatively or to a purpose, as will be suggested in Item 15: *Misprints*.

Are they pleasant to look at?

Attractive setting out will encourage a student to start working on your handout, an unpleasant page will discourage her.

Resist the temptation to fit your course onto the back of a postage stamp. Nothing is more off-putting than a page crammed full of symbols.

Space out the equations so that each one is easy to see as a complete entity. If you run equations from one line to the next the student cannot appreciate the form of the equation. It is also difficult to see what has changed and what has not changed from one line to the next.

Are they easy to read and study from?

Students will find your handouts more readable if you

a   clearly define all the symbols you use, especially details like upper and lower case letters

b   indicate where you have omitted algebraic steps

c   break up long theorems or examples into distinct steps.

They will find it easier to study from your handouts if you staple all the pages into a little booklet. That way each student is certain to have a complete set of notes.

Misprints 15

A mathematical author once jokingly remarked that the number of misprints in a mathematics book is a countably infinite set because however many you have already found, someone will always come along and find another.

Most students are reluctant to admit that they find something in your handouts inconsistent. Rather they will think that they have missed the point or that they are plain ignorant. Conscientious students may spend hours of valuable time trying to reconcile a wrong answer. Of course students require a certain amount of confidence in their own abilities before they will be prepared to question what is presented to them in black and white. The ability to read critically and not accept a statement unless it is consistent with one's own concepts needs to be developed as students become more confident.

You should own up to genuine errors however painful this may be for you. It is a very poor model for students to observe you desperately seeking to bluff your way out of an obvious mistake when the opportunity to acknowledge and correct a genuine error presents itself positively.

It is quite permissible for you openly to introduce errors or misprints into your handouts or in the text you use in class, for a specific purpose and to challenge the class to find them. This can be especially powerful with principles, rules or formulae which it is critical that the students need to know. Depending on the degree of sophistication of the class, you can structure the degree of difficulty of the error, for example by progressively revealing information such as which page or paragraph contains the error or errors, or which area of a proof may be suspect. You could also offer a real or imaginary prize to the student who finds a certain misprint first, or for the most sophisticated method students have devised for locating the error.

Some students will become over-zealous in this situation and start to question everything without giving it serious thought. Asking them to explain their difficulty in detail should quickly solve this problem.

Conducting tutorial classes

16 Scheduling tutorial classes

17 Tutorial rooms

18 Individual tutorials

19 Blackboard tutorials

20 Group tutorials

21 Student presentations

22 Right or wrong

23 No hands

24 Review tutorial

25 Training tutors

Conducting tutorials

Tutorial classes in mathematics usually take the following four forms with the first kind by far the most common.

1   The tutor walks about the room answering individual questions as students raise their hands, as in Item 18: *Individual tutorials.*

2   The tutor works problems at the blackboard, as in Item 19: *Blackboard tutorials.*

3   Students work problems in groups and the tutor moves from group to group, as in Item 20: *Group tutorials.*

4   Students give presentations, as in Item 21: *Student presentations.*

Whatever form the tutorials take, students will learn more if they come prepared. Tell the students exactly what you would like them to do to prepare for your tutorials. This may be reading through lecture notes, hand-outs or texts, attempting specified exercises or taking a computer test. Make it obvious that those students who do prepare are the ones who are benefitting from the tutorials. Many students make poor use of tutorials. They spend the tutorial hour reading through notes or looking at exercises for the first time. Indicate to such students right at the beginning of the course that these activities are not appropriate during tutorials.

Scheduling tutorial classes 16

The timing of tutorial classes can have an effect on their usefulness. If there is some flexibility in scheduling tutorial classes it is better to have them at a separate time from the lecture or just before it. Just before is probably best of all because in this way your students can:

a   take a whole week to review the last lecture, try some exercises, and come to the tutorial with questions to ask;

b   clear up any difficulties before facing up to new material in the lecture to follow, as the sequential nature of maths courses makes this very important;

c   get into a mathematical frame of mind and so be ready to appreciate the lecture.

If the tutorial class follows on immediately after the lecture students will not have time to read their notes and digest the material. Capable students may not need to do this but weak students will spend the entire tutorial hour trying to make sense of the lecture notes. They will gain no benefit from having spent the hour with a tutor and fellow students. Other students will ask questions which amount to having the lecture delivered again. It is very important for students to read through their lecture notes soon after a lecture but a tutorial class is not the place for this activity. If the tutorial class is scheduled straight after the lecture you may find yourself continually repeating the lecture. It is better if the tutorial classes lag a week behind the lectures.

You may have your mathematics classes into blocks of 3 to 4 hours. In this case you can still run a tutorial class first and follow on with lectures and other activities.

If the tutorial class is sandwiched between two lectures the students will use it as a coffee break. You will have to work very hard to keep their attention.

Try to organise a formal break of fixed length.

Tutorial rooms 17

The sizes of rooms and the way they are furnished can have an important effect on the success of tutorial classes. The room set-up generally reflects the attitude your department has to the organisation of teaching and learning in general and tutorials in particular. You may be able to rearrange the furniture to facilitate the kinds of activities you want to encourage, but this may make you unpopular with other staff and janitors. If you try to run a tutorial class in which students are supposed to work in small groups but all the seats are fixed facing the front it won't work.

Some room layouts for different kinds of tutorials are shown below.

a   If you want the whole group to take part in discussions they have to be able to look at each other.

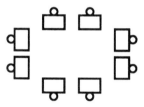

b   If you want students to work in groups it is useful to have a spare chair at each table so that the tutor can join each group.

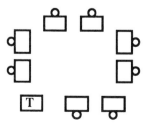

c   If you are going to use a room for several activities such as a lecture, individual work and group work you may need an arrangement like this.

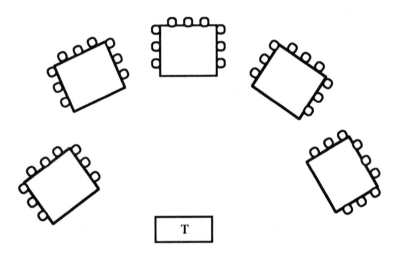

Whatever you do with the furniture it is important to note that many students have been trained at school to do maths by themselves and not to talk in class. If you want all the students to change their behaviour in maths classes you will have to do more than shift tables and chairs about. The designers of the National Curriculum in the UK have urged that exploratory talk should be an essential part of any class so future students in UK higher education may expect to talk constructively in their mathematics classes.

Room sizes also restrict the kinds of tutorials that are possible. In small rooms you will have one tutor per room while in large rooms there are often several tutors wandering about. If you have your own small room you can operate in your own individual way and also get to know the students. In a large room students have the option of choosing a tutor whom they find most useful. However, the atmosphere tends to be very impersonal.

Individual tutorials 18

The students are working individually on set exercises and usually raise their hands if they want assistance from a tutor. Here are some useful do's and don'ts.

a   Treat the students in a friendly manner, making sure they understand that you are there to help them.

b   Do not let one or two keen students use up all your time, even if no-one else seems to be clamouring for attention.

c   Shy students will find it easier to ask you a question if you are near them, so move about or stand near students whom you know to be shy.

d   If you see a student in difficulty, ask her if you could help.

e   Sit down beside the student you are helping rather than lean over them. Sitting is less threatening for the student.

f   When a student asks for help with a question, ask how he or she has approached the problem. It is better to try to proceed from the student's approach to the problem than to introduce a completely foreign approach straight away.

g   Never start writing out a solution until you are sure that you and your student are on the same wavelength. You must agree what the problem means and what method will be used before proceeding.

h   Insist that students bring their notes or text so that you can refer them to facts and examples. It is a waste of time to repeat what the student already has available in notes or handouts. This also allows you to see how well the student has recorded and appreciated what you and our colleagues have conveyed in lectures.

Blackboard tutorials   19

In blackboard tutorials the students are usually sitting facing the blackboard or OHP screen while the tutor works exercises from a tutorial sheet which the students may or may not have been given in advance. There is a danger with this kind of tutorial that students will merely try to mimic the solutions you produce without understanding the processes involved. Many of the following suggestions aim to prevent this.

a   Choose some of the problems from the tutorial sheet yourself, selecting those which illustrate basic concepts.

b   Let students choose most of the problems. They should feel that they have some control over proceedings and can get their individual difficulties attended to.

c   Allow students to interrupt the solution to ask for clarification rather than saying, *"Any questions ? "* at the end.

d   Suggest that students follow your solution, listening to what you say as well as observing what you write, and that they copy it down afterwards. Better still, write the solutions on transparencies in class and give students copies afterwards.

e   Try to explain not only what you are doing but why you are doing it, otherwise students will not be aware that you are continually making decisions on how to proceed. They tend to see a solution as an "off the rack" finished product rather than as a sequence of decisions and their implementation.

f   Instead of always providing the solution yourself, ask the students to work the solution step by step, giving prompts when necessary. Then write a solution on the board from students' contributions. While students are attempting parts of the solution, have a look what they are doing. Acknowledge those who are on the right track. Suggest a change of direction to those who have made an unpromising start. Try to get some action out of those who have made no start at all! Use the information you get from looking at students attempts to ask students to contribute to the blackboard solution.

g   Ask students to take your place as 'demonstrator'. Students are often more willing to offer suggestions to one of their colleagues and to query what they say. You can yourself help the demonstrator with discreet questions or comments if absolutely necessary.

Group tutorials   **2 0**

Group tutorials involve the students in sitting around tables in groups of three or four with an extra place at each table for the tutor. You may want to organise the groups in some way, alphabetically or with a capable student in each group. Otherwise let the students form their own groups, although you may have to help in incorporating loners into the groups. Students whose first language is not the language of instruction often have difficulty in integrating into a group: they too may need help.

At the beginning of the tutorial session students are given a sheet of exercises to work during the tutorial. If you give the exercises out in advance some students will do the exercises before they arrive, but most students will not. Groups do not work effectively if each member is trying to do something different. If necessary give students two sets of exercises, one for individual work and one for the group tutorial. The tutor moves from one group to another, sitting at the spare place and discussing progress with the group. The more effectively the group is working, the less there is for the tutor to do.

If your students are not accustomed to working through mathematics exercises in groups you will have to encourage them if they are to take an effective part in such activities. They must learn to appreciate the benefit to be gained from this kind of tutorial.

Within the groups students can:

a   agree a strategy for teaching the set problem

b   discuss and evaluate different approaches to a problem before starting on the calculations

c   check intermediate results before proceeding

d   ask for help when stuck

e   check their work for errors, doing numerical work on a calculator as necessary

f   justify what they have written if there is a difference of opinion

g   take turns to lead the group or to work on the problem with the support of the others.

As the tutor, firmly resist the urge to take control of a group or the whole class. When you join a group find out what progress has been made, elicit suggestions for the next step, give prompts, but don't "tell it all". Direct questions at a quiet member of the group or, if the quiet person has written something useful, ask him or her to explain it to the others. Use the activities in the groups to make students aware that:

- by explaining something to someone else they will often clarify their own ideas;

- by checking each step in a long problem they can save valuable time and avoid setting out on wild goose chases.

Both you and your students will need to practise if you are to learn to use this kind of tutorial effectively. The benefits are worth the effort.

Student presentations 21

Student presentations are used mainly with more advanced students. They may be oral presentations of assignments, summaries of chapters of a text or solutions to individual problems. Explain to students that oral presentations and subsequent criticism and discussion are an important part of many technical jobs: they are things in which their future employers will expect them to have some proficiency.

There are also more immediate benefits for students from making presentations. In order to explain a topic to an audience they have to think it through very carefully and prepare it thoroughly. The very act of presenting the material to others will often deepen their understanding. As well as presenting their prepared material, students must answer questions from the audience and accept criticism of their work. Some students may find it easier to ask questions of a fellow student than a lecturer.

If your students have had little previous experience at oral presentation (which is often the case with senior mathematics students) some general instruction will need to be provided. You can give this in class or give students a handout or a reference to suitable material.

Your instructions will have to cover at least the following points.

    a   Preparation of transparencies and/or use of the blackboard.

    b   Speaking clearly.

    c   Explaining results as well as writing them down.

    d   A simple format such as introduction, aim, results, conclusion.

    e   Making proper use of the allocated time.

Presumably these are all things you have had considerable experience with, s it shouldn't be difficult for you to give students directions which will mak their presentations effective. In any case, they have plenty of role model including yourself, to follow.

Right or wrong 22

One reason why students find it difficult to talk about maths problems with a tutor or with each other is because they think everything in maths is either right or wrong. They feel stupid if they are wrong so they prefer to play safe and say nothing. This attitude is the result of years of training with exercises which have 'correct' answers at the back of the book. Teachers who ridicule students when they give imperfect answers do not help either. How can you encourage more students to ask questions and participate in discussion?

a   Never make fun of or put down a student who gives an incorrect response.

b   However wrong an answer is, there is always something that can be salvaged from it. It may be the result of a common misconception, which gives you the opportunity to discuss that. It may be a special case, so you can suggest that students think of a more general answer along the same lines. It may be the opposite of the required response, which gives you the chance to clarify the concept which has been misunderstood. If one person in a class has an idea back to front there will probably be others in the same position.

c   Acknowledge the person who gave an answer in further discussion to let them know that they have made a useful contribution. You might say: *Kathy's answer satisfies all the basic conditions but there are other possibilities as well. Can you suggest some others or something that will cover all the possibilities ?*

d   Ask the person to assess their own solution, asking questions which will guide them to recognise its incompleteness or some other deficiency. You can also ask the group to spot problems. Do this also when adequate solutions are provided.

No hands! 23

It is easy when students comes for help with problems, for you to reach at once for paper and pencil. You can see at a glance where they have gone astray. There is a great temptation to correct the error, ask them if all is now understood and send them on their way.

But do students really understand what you have done? Many a student will say *"Yes, I do understand"* when the truth is closer to *"No I don't"*. How can you tell? One way is to resist the temptation to do the writing yourself. Don't pick up a pencil: no hands! Discuss, explain, point; but ask the student to do the writing. If the student really understands your explanation he will be able to write out the solution in front of you. If he can't write it out or writes it incorrectly you have some more explaining to do.

The explaining process will take longer this way because you can write the solution out much more quickly than your students. On the other hand, there is a much better chance that you will send students on their way feeling that they really understand.

Review tutorial 24

It is important to allow some contact hours to review course work. At the end of the year you may give students sets of revision exercises. Alternatively you may work questions from past papers for your students in class. If you choose the review problems you probably have difficulty in deciding how close to the examination problems they should be. You can avoid this awkward decision by involving your students more directly in the review tutorials.

Ask each student to bring along one or more questions which they find difficult to understand, especially ones that have been bothering them for some time. Most students will have quite a collection of such questions.

You may want to collect the questions before the tutorial to check that they are suitable or just to make copies of them. If you use group tutorials, ask each group to work together on the questions submitted by group members. Of course you will help the groups in the usual way. When a group has a solution to a question one member can present the solution to the whole class. If you don't use groups, you could ask your students to work individually or in pairs on the submitted questions and ask individuals to explain their solutions to the class.

If you don't have much confidence in your students you can always work the problems the students have submitted on the blackboard.

Training tutors   2 5

If you are responsible for a large class and have several tutors assigned to look after the tutorials you can help tutors and students by careful planning. Of course you will have to assign tutors and students to rooms and time-slots. There are many other things you can do to make tutorials run smoothly and in the manner you intend.

Explain to tutors exactly what you would like them to do in your tutorials. Should they keep records of attendance, provide answers or solutions, give special attention to certain students and so on. However precise your instructions are, some tutors will carry them out more successfully than others. This, in turn, will make some students feel they are being disadvantaged.

One way to help tutors conduct the tutorials in the way you envisage is to run a workshop for tutors at the beginning of the course. Depending on what kind of tutorials you have in mind you could make use of the material in Item 18: *Individual tutorials,* Item 19: *Blackboard tutorials* or Item 20: *Group tutorials* as a basis for the workshop. You could also provide them with a copy of this book!

Alternatively, show the tutors what you would like them to do by taking a tutorial with each tutor's group while the tutor observes you in action, or by inviting them into your own tutorials. You can easily make some excuse for your intrusion, such as that you are trying to get to know all the students or that you want to see how they are managing. This is very effective if you have some inexperienced tutors, graduate students for example. A real live demonstration will have much more impact than a list of instructions .

Using exercises

Exercises

Working through exercises is the universally accepted method of learning mathematics. The first item suggests improvements to standard sets of exercises. Exercises are then compared to problems.

Some alternative kinds of questions are proposed to deal with special difficulties.

Finally there are some notes on how answers and solutions can be used in the learning process.

Exercises 26

When you are making up a set of exercises ask yourself why each question should be included. Perhaps you could use a checklist to help you to decide which exercises are most suitable for your intended aims. Here is a list of some of the reasons why you might set an exercise.

- To provide practice in the application of a rule or theorem.

- To test whether several principles can be combined.

- To test whether students can discriminate between different situations.

- To lead to a new result.

- To show students an interesting application.

- To build on topics studied earlier.

- To prepare the ground for future topics.

You may also want to consider these more general points about sets of exercises.

a   Make the first exercise or two a simple direct application of the new material. It is important to get students started and to give them confidence.

b   Since sets of exercises in texts are not always well-graded you need to work through enough of each exercise to be aware of any difficulties involved. For example, does the question turn into a very nasty integral?

c   If discrimination between different procedures is important then set some exercises in which students can concentrate on this. Longer questions on the same topic can follow.

d   Make sure you have exercises which illustrate all the implications of a rule and common misconceptions about it. Item 31: *Counter examples* explains a useful way to do this.

e   Some students need more exercises than others to master new concepts. If you set very long lists of exercises, give students some advice on how they should approach them, e.g. *"If you had no trouble with Exercises 1 - 4 then skip Exercises 5 - 9"*.

f   Try to avoid making a set of exercises boring and repetitive. If students can do your exercises quite mechanically they will stop thinking and not derive the intended benefit from their work.

Problems 27

Some people make a distinction between exercises and problems. Exercises are supposed to be routine applications of rules while problems are more demanding and difficult. But as Schoenfeld points out:

> *The difficulty with defining the term problem is that problem solving is relative. The same tasks that call for significant efforts from some students may well be routine exercises for others, and answering them may just be a matter of recall for a given mathematician. Thus being a problem is not a property inherent in a mathematical task. Rather, it is a particular relationship between the individual and the task that makes the task a problem for that person.*

For a large proportion of students who are studying mathematics as a service subject the distinction between problems and exercises can be made more specific. An exercise is something for which the student has been provided with a model solution, while for a problem he has no specific model at hand. Of course one can still argue about how specific the model must be! It is important to be aware of this situation when constructing sets of questions for your students to work on.

Unless you specifically require your students to develop problem solving skills it is safer to restrict your questions to slight variants of given examples. Even with these many students experience great difficulty.

If you really want your class to develop mathematical problem solving skills then you will have to give some specific attention to developing these skills. Ideas for this can be found in Item 28: *Problem solving*.

Reference

Schoenfeld, A.H.   *Mathematical Problem Solving*, Academic Press. 1985.

Problem-solving 28

A great deal has been written about the importance of students being able to apply their mathematical knowledge to solving problems. There is much less in the literature on what problem solving skills actually are and how they can successfully be taught.

Schoenfeld has provided very detailed protocols of how students attempt to solve problems and characterised mathematical problem solving performance according to the four headings below.

Resources

Mathematical knowledge possessed by the individual that can be brought to bear on the problem at hand.

Heuristics

Strategies and techniques for making progress on unfamiliar or non-standard problems; rules of thumb for effective problem solving.

Control

Global decisions regarding the selection and implementation of resources and strategies.

Belief Systems

One's 'mathematical world view' - the set of (not necessarily conscious) determinants of an individual's behaviour.

He has shown that if you wish to teach heuristics then you must do it in a very detailed way. General instructions such as *"Try special cases"* are not very helpful and need to be replaced by numerous sub-strategies, such as those listed below.

Strategy 1

If there is an integer parameter n in a problem statement, it might be appropriate to calculate the special cases when n = 1, 2, 3, 4 (and maybe a few more). One may see a pattern that suggests an answer, which can be verified by induction. The calculations themselves may suggest the inductive mechanism.

Strategy 2

One can gain insight into questions about the roots of complicated algebraic expressions by choosing as special cases those expressions whose roots are easy to keep track of (e.g. easily factored polynomials with integer roots).

Strategy 3

In iterated computations or recursions, substituting the particular values of 0 (unless it causes loss of generality) and/or 1 often allows one to see patterns. Such special cases allow one to observe regularities that might otherwise be obscured by a morass of symbols.

Strategy 4

When dealing with geometric figures, one should first examine the special cases that have minimal complexity. Consider regular polygons, for example;or isosceles or right or equilateral rather than general triangles; or semi- or quarter-circles rather than arbitrary sectors, and so forth.

Strategy 5

For geometric arguments, convenient values for computation can often be chosen without loss of generality (e.g. setting the radius of an arbitrary circle to be 1). Such special cases make subsequent computations much easier.

Strategy 6

Calculating (or when easier, approximating) values over a range of cases may suggest the nature of an extremum, which, once thus determined, may be justified in any of a variety of ways. Special cases of symmetric objects are often prime candidates for examination.

This classification may not be exhaustive but does show the kind of detail that is required if students are going to learn to use such strategies.

To make use of lists of strategies like this students obviously need large numbers of problems on which to practise until they develop a feel for which approach is appropriate for which kind of problem. It is a further leap forward to devise strategies for completely new kinds of problems.

If you think that developing problem solving skills is an important part of mathematical education then you and your students must be prepared to devote a great deal of time to this activity.

A good way to find out how your students approach problems is to ask some of them individually to think aloud into a tape recorder as they work on a problem or; ask pairs of students to work together on a problem and record the discussion. You can use the tapes yourself to find out about the problem solving skills of your students. You could also play suitable tapes during a tutorial in order to:

a   help your students to understand what problem solving is all about;

b   make students aware of the strategies being used or misused;

c   let students witness the effects of errors of judgment and suggest remedies;

d   encourage students to reflect on their own thought processes as they attempt to solve problems.

Problem solving has not been treated as a distinct part of university mathematics courses in the past. However, as computer packages take over the solution of ever increasing numbers of routine problems, the theory and practice of problem solving may become the most important aspect of the mathematics that we teach.

Reference

Schoenfeld, A.H., *Mathematical Problem Solving*, Academic Press, 1985.

Make up your own problem 29

As an occasional alternative to handing out the usual weekly set of exercises ask students in a tutorial to invent some problems of their own. This is a good strategy in statistics, where students learn to test hypotheses in a variety of situations, though if too many situations are presented at one time students may not get enough opportunity to distinguish one situation from another. While this activity may be more difficult in other branches of mathematics, it may be well worth your while to try it out since important psychological benefits are to be gained from this process. For example, a colleague in Bristol Polytechnic asked students to make up their own problems to do with the theory of structures. He claimed that the single most important gain was that, when students ran into difficulties with the mathematics, they asked for help in solving their problem rather than his.

Ask the students to make up problems from other subjects they are studying, from things that may have cropped up at work, from their hobbies or from articles they have seen in the press. This encourages students to see how mathematics and statistics relate to their own lives. You may wish to focus the problems a bit, for example by nominating a particular type of hypothesis, say, a difference of two proportions. Don't be surprised when the first suggestions are about differences of means or correlation.

When the group is satisfied that they have an appropriate problem to investigate, ask them to contribute some imaginary data. The data will probably give a strange value of the test statistic. This provides a good opportunity to discuss what properties of the data caused the strange result. It can also help to make previously forbidding formulae come to life. When one's own data are substituted into a formula it takes on a real meaning.

If you give students short tests, you could incorporate a *"make up your own problem and solve it"* item. In this case it is necessary to give some definite instructions about the amount of imaginary data required and what the students should do with it. Here is an example of such a question.

Make up a problem about your leisure activities, sports, etc. which could be solved by calculating a correlation co-efficient. Invent 6 data points, test the significance of your co-efficient and comment on the result.

Open-ended questions 30

There is a strong tradition only to give students problems to which there are unique correct answers. However, this is quite atypical of real world mathematical problems, which often have many possible solutions.

One way to make traditional questions more realistic is simply to omit some of the information that is usually given. Here is an example.

The table below classifies the last 70 cases brought before a juvenile court according to the size of the family from which the young offender came.

Number of children in family	Number of cases
1	3
2	3
3	9
4	16
5	8
6	15
More than 6	16

Can you use this information to justify the claim that juvenile offenders mostly come from large families? Is there any other information you would like to have in order to be able to draw a conclusion?

This is part of a problem from a book by Anderson and Loynes which contains many other open-ended statistical problems.

As with other innovations you will need to explain to students why you are asking non-traditional questions. Making problems more realistic is quite easy to justify.

At first students may suggest that the questions are impossible to answer as they may never have encountered open-ended questions before. They may even have been taught that unless every last detail has been specified the question cannot be answered. It may take a little time for this new approach to be accepted.

As well as being more realistic, open-ended problems are a good device for starting discussion, a difficult thing to do in maths classes, as discussed in Item 22: *Right or wrong*.

Reference

Anderson, C.W. &   *The Teaching of Practical Statistics.*
Loynes, R.M.   Wiley, 1987

Counter-examples   31

Counter-examples are a very powerful tool in the development of new concepts and the learning of new rules. They are far too little used for this purpose. Have you ever seen a set of quadratic equations to be solved which contains:

$$x^3 - 5x + 6 = 0 ?$$

We tend to throw up our hands in horror when students do not discriminate correctly, yet we seldom give them opportunities to develop the necessary skills.

Each time students are presented with a new concept it is vital that they understand exactly what is involved and what is not. Counter-examples can help the student to make a fuzzy concept precise.

Many theorems require several conditions to be satisfied before they can be applied. A good way to check that your students are paying attention to all these conditions is to include counter-examples in the exercises, examples to which the theorem should not be applied. For example, when showing students how to apply Leibniz' test for alternating series, you could ask students whether the test is satisfied by the following series and if not, why not.

(i) $\dfrac{1}{2} - \dfrac{1}{4} + \dfrac{1}{3} - \dfrac{1}{5} + \dfrac{1}{6} - \dfrac{1}{7} + \ldots\ldots$

(ii) $\dfrac{1}{2} - \dfrac{2}{3} + \dfrac{3}{4} - \dfrac{4}{5} + \dfrac{5}{6} - \ldots\ldots$

Making use of student errors 32

You have probably been concerned many times at the dreadful mistakes in elementary algebra made by your students. Students tend to make these errors when their attention is focussed on mastering some new technique and not on the algebra.

Instead of complaining to your colleagues about such errors, collect them. After a few weeks, perhaps before a major test or examination, prepare a set of exercises based on your collection. A simple way to do this is to make a list of pairs of expressions such as:

$$\sqrt{x^2 + 4} \ , \ x + 2$$

$$\tan^{-1} \sqrt{x} \ , \ \frac{1}{\tan \sqrt{x}}$$

Slip in some pairs that are equal, but look similar to pairs that are not equal, and ask students to identify the pairs as equal or not equal. Ask them to justify their decisions, because this will enable them to understand why they made the errors.

This is a good exercise for students to work on in pairs.

Depending on your relationship with the class, you may want to acknowledge the students whose work contributed to your list.

Answers 33

By providing answers with exercises you enable students to find out whether they are perfectly correct or not. Students certainly want to know this but an answer on its own is not a great aid for error detection. Also, providing a single answer is often misleading because there may be other "correct" forms of an answer. Students can get as much value as possible from a set of exercises with answers if you suggest that when they do not agree with the given answer they should, in approximately the following order:

•   try to transform their answer into the given one

•   check their work for errors

•   compare solutions with one or two other students. This may suggest that the given answer is not correct.

•   work back to see how they could have reached the 'correct' answer

•   seek help from a tutor.

The first of these suggestions is a very good exercise in itself. It often requires a good knowledge of identities and considerable strategy successfully to change the form of an answer. Practice in obtaining a given form of the answer is valuable for answering examination questions which commence with: *Show that* ....

By providing solutions or outlines of solutions instead of just answers your students may be able to help themselves more readily.

Using solutions 34

There is a certain amount of resistance to providing students with solutions to exercises in advance. Common remarks are: *"If you give them the solutions they won't do the exercises for themselves "* or *"They will just memorise the solutions."* This need not happen.

Firstly, students are not really so stupid as to think that merely possessing a piece of paper with solutions on it is equivalent to understanding how to do the exercises. If they do think that, your immediate task is clear!

Secondly, weaker students may be encouraged to try exercises which they might otherwise find too daunting when they have the solutions to fall back on. Being unable to solve any of the questions on a tutorial sheet can be a very deflating experience for a student.

Thirdly, if you don't think your students know how to use solutions sensibly then they need some guidance from you.

Finally, students only memorise solutions as a last resort. If they gain some understanding by studying your solutions they won't need to memorise them.

Solutions can be provided in many different forms, for example, your annotations alongside a full solution, printed off and made available to those who want them, would satisfy most students. The type of solutions you provide will depend on how much help you think your students need. John Searl at Edinburgh University provides solutions in two forms. Brief written solutions and, for students who find this insufficient, a detailed explanation on video which can be viewed in the library.

Overleaf is an example of two solutions to the same question, the first brief, the second very detailed, and, on page 99, a brief set of instructions on the use of solutions.

Question

Find the general solution of the equation $(1 + x^2)(dy - dx) = 2xy\ dx$.

Solution 1

$$\frac{dy}{dx} - \frac{2x}{1 + x^2}\ y = 1, \quad \text{linear}$$

$$\exp\left(\int \frac{-2x\ dx}{1 + x^2}\right) = \frac{1}{1 + x^2}$$

$$\frac{y}{1 + x^2} = \tan^{-1} x + c$$

$$\text{so } y = (1 + x^2)\tan^{-1} x + c(1 + x^2)$$

Solution 2

The variables are not separable and the factor $(1 + x^2)$ rules out the homogeneous type so try to write the equation in the form:

$$\frac{dy}{dx} + P(x)\ y = Q(x)$$

$$\frac{dy}{dx} - \frac{2x}{1 + x^2}\ y = 1 \qquad \left[\begin{array}{l}\text{Divide the equation by} \\ (1 + x^2)\ dx \ \text{and transpose}\end{array}\right]$$

The integrating factor $\rho(x)$ is

$$\exp\left(\int \frac{-2x\ dx}{1 + x^2}\right) \qquad \left[\begin{array}{l}\text{The integrating factor is} \\ \exp\left(\int P(x)\ dx\right)\end{array}\right]$$

$$= \exp\left[-\ln(1 + x^2)\right] \qquad \left[\text{note that } \frac{d}{dx}(1 + x^2) = 2x\right]$$

$$= \exp\left[\ln(1 + x^2)^{-1}\right] \qquad \left[\begin{array}{l}\text{note that } -\ln u = -1 \ln u \\ \qquad\qquad\quad = \ln u^{-1}\end{array}\right]$$

$$= \frac{1}{1 + x^2}$$

[note that $\exp(\ln u) = u$]

So $\dfrac{y}{1 + x^2} = \displaystyle\int \dfrac{1}{1 + x^2}\, dx$

$$\left[y\rho(x) = \int Q(x)\,\rho(x)\, dx \right]$$

$$= \tan^{-1} x + C$$

[standard integral]

So $y = (1 + x^2)\,\tan^{-1} x + c\,(1 + x^2)$

Here is a brief set of instructions which you might want to offer your students:

What to Do With Solutions

1   If you can't get started on a problem have a look at the first step in the solution and then try to proceed on your own.

2   If you get bogged down in an algebraic expression, glance at the solution to see if it should be complicated or not.

3   When you arrive at an intermediate result, check it before proceeding, this can save time and encourage you to go on.

4   Use the solutions to correct your errors. Next time you work the same problem or a similar one try to do it without relying on the solution.

Don't forget that most of your students are not even fledgling mathematicians. They have to learn to crawl before they can walk and they will probably never fly! All in all, solutions can save students a lot of time and actually increase the number of different problems that they attempt.

Simplified answers 35

Until very recently an important part of mathematical activity was finding formulas which could be computed using paper, pencil and various tables. This was an essential objective of mathematical research and should not be underestimated. The accessibility of computers and calculators has dramatically changed this situation. Many of the very elegant methods of transforming formulae to make them easier to compute are superfluous when every student has a pocket calculator.

The most common instruction even in the school texts of today is *Simplify the following*. The purpose of simplifying expressions was to write them in the form from which it was as easy as possible to calculate a numerical answer. Since the most cumbersome arithmetical operations were division and finding square roots, the number of these operations had to be minimised. Hence we simplify:

$$\frac{1}{f} + \frac{1}{g} + \frac{1}{h} \quad \text{as} \quad \frac{gh + fh + fg}{fgh}$$

There is now only one division, but there are more operations and the result clearly looks less simple than the original. Other examples are the removal of surds from the denominator so that:

$$\frac{1}{\sqrt{3} + \sqrt{2}} \quad \text{becomes} \quad \sqrt{3} - \sqrt{2} \ .$$

There are still situations in mathematics when operations such as those above are desirable or even necessary. For example, it is useful to be able to recognise that the two surdic expressions above are equal. There are also many situations where such operations serve no useful purpose.

When you set exercises in calculus and provide the students with answers, your students will find the answers much more helpful if you leave them unsimplified. You may be so used to simplifying answers, because that was part of your training, that you are hardly aware that you are doing it. You need to ask yourself as you work each exercise: *"Is there any point in going any further? Is there any point in wasting students' time doing transformations which do not simplify calculations?"*

If you leave the answer unsimplified the student will know at once whether he has performed the main object of the exercise correctly or not. In spite of this many mathematicians think it is good for their students to have to struggle to obtain the given answer. Students on the other hand find it frustrating and pointless. Perhaps if Newton and Leibniz had owned pocket calculators they would have written their results differently!

Developing study skills

Study skills

Students are frequently advised when studying mathematics to *work through all the given exercises and then work through them again.* However, there is more to the study of mathematics than this simple instruction implies. In this section many of the difficulties that students experience while studying mathematics are identified. Many of these difficulties are caused by the conventional language and symbolic nature of mathematics. Numerous activities are suggested to overcome the problems that have been identified.

Questions   36

A good teacher asks questions which act as prompts to direct students into thinking along certain paths. Polya's questions in *How to Solve It* are excellent examples. But most students do not spend their days with a teacher like Polya. They listen to lectures in large groups and have to do most of their learning alone. Therefore it would be very helpful if students could ask themselves questions as they study mathematics. This self-questioning is a standard procedure in some programs for improving reading and comprehension such as SQ3R (a reading method: Survey, Question, Read, Recall, Review). It may not be feasible to formalise the questions as in SQ3R in a mathematical context, but if students can get into the habit of asking themselves questions as they study they may find it helpful.

Here are some very simple questions with general application.

- What does this symbol mean?

- What is the difference between the capital X and the lower case x?

- What has been done to a symbolic expression to turn it into the one which follows it?

- Where did that piece of information come from?

Sometimes the student will be able to find answers to her questions herself, sometimes the very act of formalising the question reveals the answer. If the student can't answer herself then at least she has a question to ask another student or tutor. The trouble is that framing the right question at the right moment is no easy task. Ask your students to listen carefully to the questions you ask, not just so that they can answer the questions but so that they can ask themselves similar questions as they study. This is a completely different way of listening to your questions. You could do this in the context of a regular lecture. Ask the students to read a few lines from a handout or text. Definitions or short proofs are suitable. Then ask some questions to test the students understanding of what they have read. Many students will improve their understanding of what has been read as a result of your questions and may gradually learn to ask the questions themselves.

For example, ask students to read the following definition.

> The n functions $f_1, f_2, ..., f_n$ are said to be linearly dependent over a interval I if and only if there exists a set of n constants $c_1, c_2, .., c$ at least one of which is different from zero, such that the equation,
>
> $$c_1 f_1(x) + c_2 f_2(x) + ... + c_n f_n(x) = 0$$
>
> holds for all values of x in I.

Then ask a series of questions to help students to understand what the definition says and to relate this new information to their existing ideas.

- Have you seen a definition very similar to this in another area mathematics? Which area?

- What does the definition look like if there are 2 functions, f_1 and f_2 ?

- Can you make up two functions that are linearly dependent or linear independent?

- Does the test work on them as expected?

- What does the test look like for three functions?

- What about $\sin x$ and $\cos x$? I know $\sin^2 x + \cos^2 x = 1$. Does that make them linearly dependent?

Reference

Polya, G.,   *How to Solve It*
Doubleday. 1957.

Reading 37

The skills that are required to read verbal and symbolic text in mathematics are very different from more general reading skills. It is obvious that if you remove a single symbol from an equation you change the whole meaning or make the equation quite meaningless. The same is true for mathematical definitions expressed in words, but many students are not aware of this. Furthermore, it is often the little words which are overlooked, words such as any, all, and at least which are crucial to the meaning of a definition. If a student has forgotten what *homogeneous* means in a certain context he can look it up in the index and find the definition, but this is not the case for the little words.

Apart from the need for attention to every detail there is the problem that mathematical text is very sparse, there is no redundancy, as explored in Item 39: *Missing words*. Standard reading techniques such as looking for key words are not very useful because in mathematics there are no redundant words or phrases: every word is a key word.

To help your students to read mathematical texts you could point out some of these difficulties to them. At least this would make them aware that their difficulties are due to the nature of mathematical writing and not their own stupidity. Explain that reading unfamiliar mathematics is slow even for mathematicians. Give an example of something that you had to read many times before you felt you understood it completely. Explain also that since mathematics is highly sequential it is difficult to read through a whole chapter, or even a section of a chapter, at once. This is because each new definition or theorem require an understanding of earlier definitions and theorems. Most people do not gain this understanding merely by reading. They need to work through some examples to consolidate all the new concepts in preparation for building on them.

You can use a class exercise to increase students' awareness of these problems. Give your class something to read, the beginning of a chapter or a short paper, something concerned with the topic they are studying so there is a good reason for them to read and understand it. After a few minutes ask them to discuss in groups of two or three what they have read and any difficulties they had with their reading. Ask also for strategies for overcoming some of these difficulties. Then have the groups report back to the whole class.

If you can make your students aware of these general ideas you may lessen their feelings of inadequacy and encourage them to persevere with what is undoubtedly a difficult task.

Words into symbols   **38**

It is well-known that mathematics students at all levels have difficulty with problems expressed in words. Some educators have tried to deal with this issue by teaching students to recognise and solve large numbers of stereotyped problems. Students taught in this way are easily confused by slight changes in the wording so this seems a rather futile approach.

The difficulties with word problems are closely associated with the more general problem of mathematical reading skills discussed in Item 37: *Reading*.

For word problems students must be able to write down what is given and what is required in appropriate symbolic form. Because the original problem is in words, students often write what is given in verbal form rather than symbolic form. For example, they will write:

$$\text{mean} = 24.3, \quad \text{s.d.} = 5.7$$

This is correct but not precise enough. It does not say whether 24.3 is the population mean or the sample mean and by labelling with the word *mean* the necessity for making the distinction may be overlooked. If there are standard symbols such as μ and x in the above example, students should use those.
If they have to choose letters for variables they need to make the choices explicit. For example in a problem involving rates, any given rate is liable to acquire the label:

$$\frac{dy}{dx}$$

regardless of what quantities it refers to. Students are not sufficiently aware of the hidden culture in mathematics in which certain letters represent special kinds of things. As a result they can read or write:

$$\frac{d}{dc}e^{cx}$$

without thinking that it looks a little strange.

You may think that because you habitually label the given data appropriately when working examples for students that they will automatically follow suit. On the contrary, unless these issues are specifically raised with the students several times some of them will simply not notice what you have been doing.

Missing words   39

Mathematicians like to express themselves as concisely as possible. They compete with each other to produce definitions which contain all the necessary information in as few words as possible. This is considered the height of mathematical elegance. This sparseness in the use of words and the absence of redundancy makes mathematics very difficult for the beginner to read. A simple exercise can demonstrate this aspect of mathematical text very clearly. Just making the students aware of the problem often helps to overcome it.

You will need two short extracts, one from, say, a newspaper and one from a mathematics book. You can remove two or three words from each sentence in the newspaper article and the meaning will still be obvious. When you do the same thing in the mathematics text the removal of even some apparently insignificant words like *all* , *some* or *each* will cause confusion.

Give the students the modified newspaper article first and when they have had time to read it ask someone to read it aloud filling the gaps with words which they consider appropriate. This will be very easy to do. (The exercise could also be done by asking students to fill in the gaps by selecting words from a list of those words which you had deleted.) Then repeat the exercise with the mathematics text. The increased difficulty of the second task will be apparent at once. This exercise should make the students aware that they must read and take notice of every word, especially every little word in mathematics. The need to think about every word in a mathematical sentence and the need to read and re-read makes mathematical reading very slow. Explain this to the students as otherwise each one of them tends to feel that she is the only person who reads mathematics slowly. Here is an example, in which missing words are indicated by the symbols - - -.

1   **Tandem parachutists dive 3000m to death.**

Melbourne. A man making his first parachute - - - and an experienced sky-diver - - - yesterday when their single chute - - - to open.

The two Melbourne men, buckled - - -, plunged to their - - - from about 3000m at Corowa on the Victorian-New South Wales border.

Spectators watched in - - - as parachuting instructor Julian Stuart Ashley Brown, 34, of Box Hill, and Goran Radajcic, 21, of Heidelberg, - - - to their deaths.

The men - - - out of a plane strapped together in a double harness. The main - - - failed to open.

2 Definition of Inverse Functions

Let $y = f(x)$ be a function, defined for - - - x in some - - -.
If, for - - - value y_1 of y, there exists - - - one value - - - of x in the interval such that $f(x_1) = y_1$, then we can define an - - - function $x = g(y) =$ the unique number x such that $y = f(x)$.

Our inverse function is defined - - - at those numbers which are - - - of f. We have the fundamental - - - $f(g(y)) = y$ and $g(f(x)) = x$.

Example 1. Consider the function $y = x^2$, which we view as being - - - only for $x \geq 0$. - - - positive number (or 0) can be written - - - as the - - - of a positive number (or 0). Hence we can - - - the inverse function, which will also be defined for $y \geq 0$, but not for $y < 0$. It is nothing but the - - - root function.

Missing lines   **40**

Just as each person needs to write down a certain amount of their mathematics to help their memory, so, when reading, each person requires a different amount of detail to be provided. Since the amount of detail provided in lectures and texts is fixed, it may not be sufficient to allow all students to follow from one line of working to the next. As the mathematics becomes more advanced increasing numbers of algebraic steps are omitted and the reader is expected to insert them him or herself.

The kinds of steps that are frequently omitted, whether singly or in combination, are:

a   removing brackets and regrouping terms;

eg. $C_1 e^{(s + ti) x} + C_2 e^{(s - ti) x} = e^{sx} (C_1 e^{itx} + C_2 e^{-itx})$

b   writing fractions with a common denominator;

eg. $1 - \dfrac{x^2 + 1}{x^2 - 1} = \dfrac{2}{1 - x^2}$

c   factoring;

eg. $\sqrt{a^2 x^2 + b^2} = a\sqrt{x^2 + (b/a)^2}$

d   multiplying by 1 in the form a/a;

eg. $\displaystyle\int \sec x dx = \int \dfrac{\sec^2 x + \sec x \tan x \, dx}{\sec x + \tan x}$

e   adding zero in the form a - a;

eg. $x^2 - 8 x + 3 = (x - 4)^2 - 13$

f   taking logarithms or anti-logarithms;

eg. $\ln |x| - \ln (1 + y^2) = c$
$\Rightarrow x = k (1 + y^2)$

If students are aware of these simple procedures they will have less difficulty inserting the missing lines.

When reading from one line to the next it also helps to check what has changed and what has not changed. You can give your students some experience in inserting missing lines of algebra by deliberately leaving out some steps in handouts, as in Item 13: *Uncompleted handouts*. Or you could do the same thing as an activity during a lecture or tutorial, using a format such as that shown below.

The factorial movement-generating function for the geometric random variable y is defined by:

$$\psi_y(t) = E(t^y)$$

$$= \sum_{k=1}^{\infty} t^k q^{k-1} p$$

$$=$$

$$=$$

$$= \frac{pt}{1 - qt} \text{ for } |t| < \frac{1}{q}.$$

So $\psi_y^{(1)}(t)$ the first derivative of $\psi_y(t)$ is

$$\psi_y^{(1)}(t) = \qquad\qquad = \frac{p}{(1 - qt)^2}$$

and $\psi_y^{(2)}(t) =$

Depending on the ability of the students you may have to give some hints at first as to what kind of steps would be useful.

Most importantly, students should insist on knowing how each line follows from previous ones and should not accept anything on your, or anyone else's authority.

Writing 41

Mathematics is done in the head, not on paper. A mathematician records just enough steps so that she does not get confused with details. The amount that is written down varies tremendously from one person to another.

Many students receive instructions at school about exactly what should be written down for certain types of exercise. This is unfortunate as these rituals are very hard to alter. Students continue year after year writing down lots of mathematics which they are quite capable of doing in their heads, where it really belongs.

Inevitably when you work exercises in your lectures you will:

a write down more steps than you, and some students, require in an effort to make every detail explicit, or

b leave out some routine procedures, forgetting the lack of sophistication of some of the students.

Remember that some students will try to copy your solutions verbatim, so it might be helpful if you were to indicate how much detail you expected them to write. You could dramatise the situation for your students by first writing out a very detailed solution. When you have finished you could say: *"Now I will do the problem as if I were doing it for myself and not for you"*.

You could then suggest that each student should find his own level somewhere between the two extremes, as in Item 34: *Using solutions*. Alternatively, you could use students' tests or assignments to exemplify what you want to say about how much should be written down.

Control   42

In order to work through a set of exercises students need to have some knowledge of the subject and notes or books for reference. They will make more efficient use of the time available for working exercises if they also exercise some control over their actions.

How many times have you seen an examination script containing line after line or even whole pages of working which is completely misdirected? The student should have realised it was all wrong and given up ! At examination time it is too late to bemoan this lack of control. It needs to be discussed when students are practising exercises.

Suppose a student is working through some differential equations and arrives at an integral which looks unfamiliar. What should she do ?

Should she:

- try to transform the integral?

- check the working so far to see if she has made an error?

- go back to the question to see if she has copied it wrongly?

- start again, using a completely different approach?

- look at the answer to see if it contains any clues?

- do several of the above?

- give up and start on a different exercise?

Even if she does consider some of these alternative actions, in what order should she proceed? She must also decide how much time to allocate to the above activities if they do not quickly lead to success.

Systematic consideration of the alternatives and sensible decisions can mean the difference between success and failure. As students are invariably short of time it is a good idea to introduce the notion of control over decision-making as a time-saving device. It is also vital for successful performance in examinations. However, students are unlikely to make good decisions in examinations if they are not accustomed to doing so when practising exercises.

A handout with suggestions on what to do when in difficulty might help your students. You can construct a handout for a particular topic, e.g. techniques of integration, or a more general handout for use with any kind of problem.

On the page opposite are some suggestions for a general handout on control.

Control

When you are working problems it is easy to waste a lot of time going off in the wrong direction or continuing with working that has got very complicated because your error has been made. Try to save time by referring to the following list as you work exercises or answer tests and examinations. As soon as you suspect something has gone wrong or if you have been stuck for 10 minutes:

- Re-read the question a few times. Does it actually say what you thought it meant at first?

- Have you copied numbers and formulae from the question correctly?

- Have you labelled the data correctly?

- Are the formulae you supplied in the right form?

- Are the formulae you supplied appropriate?

- Does your diagram follow from the given information?

- Have you checked for algebraic mistakes such as minus signs in front of brackets?

- Have you tried starting the problem again without looking at your earlier effort?

- If the answer is given, have you looked at it to see if it contains any clues?

Not all these questions will be required for each problem. Each question should only take a short time to answer. If none of the above suggestions leads to a solution, do not waste any more time. Go on to the next question. Seek help later if possible.

Learning through diagrams 43

Many students find it easier to recall a theorem or definition in pictures rather than words. The diagram may focus the students attention on the meaning of a theorem. Suitable diagrams are generally provided in texts and notes, but this does it mean that students will make good use of them in their learning. Here is an example of part of a session which aims to improve students' use of diagrams.

Give the students the following statement.

> *If a function f is continuous on a closed interval [a, b] and differentiable on the open interval (a, b), then there exists a number c in (a, b), such that*

$$\frac{f(b) - f(a)}{b - a} = f(c)$$

The lecturer's part of the discussion, depending of course on the students' responses, might take the following form.

> *Read the statement of the theorem again.*
>
> *Are you sure you know the precise meanings of all the words and symbols?*
>
> *Kathy, what do you understand by "a function is continuous on [a, b]"?*
>
> *John, what is the formal definition of continuity on an interval?*
>
> *Sally, why are there square brackets around the first a, b and then round brackets?*
>
> *All of you, draw a diagram of any function which is continuous on [a, b] and differentiable on (a, b).*

Peter, what does $\dfrac{f(b) - f(a)}{b - a}$ represent in your diagram?

All of you, complete the diagram to show what the theorem says.
Does looking again at the proof of the theorem make you change your mind about your diagram?

Look at your diagram and, without looking at the statement of the theorem, try to make up the theorem from the picture.

Did anyone draw a diagram in which there is more than one value of c?

Draw a diagram in which there are exactly two values of c.

Can you draw a diagram of a function which is not continuous or not differentiable and show how the theorem breaks down?

By the use of suitable questions the student is directed to relate the statement of the theorem to the diagram and vice versa. She should also become aware of which aspects of the diagram are arbitrary and which are not. This is not always clear when just one diagram is provided.

Graphs 44

When you suggest that your students draw a graph of a function or an equation, most of them will either:

a   start calculating (x, y) pairs, plotting them and joining them up somehow, or

b   set about differentiating to find maxima or minima.

Very few students will look critically at the function, recognise parts of it and organise this knowledge to sketch the curve.

It is fairly obvious why students respond with (a) or (b). These techniques correspond to exercises they have done many times. The kind of response you would like - a sketch based on the important properties of the function - is not usually part of any syllabus. As students learn each new kind of function they are shown its graph, but they don't actually do anything with the graphs themselves. The ready availability of graphics packages, which are well suited to the development of a familiarity with graphs of all kinds, may fill this gap. The effect is not yet evident among undergraduates.

As with other suggestions in this book, if you want your students to do something, you have to show them and make it possible for them to do it themselves. Whenever an opportunity arises, ask them to sketch a graph before they solve a problem analytically.

Graphs of simple functions can be used both for finding initial values to use in the numerical solution of equations and for interpreting their points of intersection. This is something that is often neglected when a numerical method is being explained.

Another point at which the graphs of functions can be utilised is in the construction of the graphs of the hyperbolic functions from the graphs of e^x and e^{-x}. Ask the students to do the construction themselves, they may never have added and subtracted ordinates in this way before.

Once students are convinced that there is some purpose in being familiar with certain standard graphs and with methods of manipulating them they will take them more seriously.

Diagrams in problem-solving 45

Once upon a time we all studied Euclidean geometry at school and so at an early age we learned about the value of diagrams to the solution of problems. Much of that geometry has now been replaced with other things so that students no longer get practice in drawing diagrams from verbal instructions. You shouldn't assume that your students will find it natural to draw a diagram as an initial step in solving a problem, even if the problem appears to cry out for a diagram. In fact it is very difficult to solve some kinds of problems, for example problems about volumes of revolution and surface area, without reference to a diagram. Frequently a good diagram will lead directly to a simple solution.

Students today tend to be obsessed with formulae. As they read a problem, they search mentally through their repertoire of formulae. They then attempt to substitute the given information into one of their 'off the rack' formulas, even if the fit is not very good. The possibility of starting with a diagram is frequently not even considered.

It is particularly difficult to alert students to the need to draw diagrams when the situation is not geometrical in any way. For example, many problems in probability can be simplified with the aid of a tree diagram.

Not only is an appropriate diagram vital to the solution of certain problems; so also are the labels that are attached to the various parts of the diagram. In calculus problems it is important to distinguish between those aspects of the diagram that remain constant and those that vary. The diagram must also be as general as possible so that the problem is not solved only for a special case. Students need to experiment with various diagrams to make sure that everything that is given is accounted, for but that no spurious restrictions have been incorporated. A very simple example is that students tend to make all triangles right-angled.

A final difficulty is that the diagrams which the student encounters in examples in texts and handouts are complete and perfect. The student does not experience the construction of the diagram.

What can you do to help your students with all these difficulties? You can certainly emphasise the importance of diagrams when you solve problems during your lectures and tutorials. Draw the diagrams in front of the students rather than presenting them ready made on a transparency. Giving the students the opportunity to see the diagram develop is more important than the neatness of the diagram itself.

Leave spaces for diagrams in uncompleted handouts so that the students have to draw them themselves.

When you set exercises and problems in which you think a diagram would be useful, give a specific instruction for a diagram to be drawn. You may think such an instruction rather childish, but you have to remember that your students may not have had the kinds of experience that you had with diagrams when they were younger.

Assessing students' work

Assessment

Mathematics has traditionally been assessed almost exclusively by written examinations containing fairly stereotyped questions.

Some modifications to standard examination procedures are suggested in this section, as well as other forms of assessment such as assignments and reports which can be used to complement or replace standard examinations.

Formulae   46

If your subject makes use of a lot of formulae and you do not provide them or allow students to provide their own you will encourage students to learn their mathematics by trying to memorise everything. There is considerable evidence that mathematics that is memorised for examinations is very quickly forgotten.

It is common practice to provide a list of formulae, standard integrals or Laplace transforms with examination papers. When you do, this students will find it helpful if you give them copies of these lists well before the examination. They can then do their examination preparation knowing exactly what information will be supplied, in what form and what they have to supply themselves. Alternatively, provide the students with a list of basic formulae and give them the opportunity to annotate and extend it. You could improve the learning process and give the students more responsibility for their own learning by asking students to construct their own formula lists. In this case you will need to specify exactly how much information they may bring into the examination. You may allow one or two A4 pages or give the students standard sized cards on which they can write their lists.

The first time they are asked to do this some students will try to encapsulate the whole course in all its detail by using very small type. However they will learn very quickly that information in this form is difficult to access quickly and therefore not very useful in an examination.

Students will produce more valuable lists if you give them some help with making the lists, as in Item 47: *Making your own formula list.*

Make up your own formula list 47

The formula list is best thought of not as a one-off task done just before the examination but as the essential part of summaries prepared during the entire course.

As with anything else you ask students to do, they will do it better if you discuss the technique with them. Ask students to make summaries of their work each week. These can contain more than formulae. Encourage students to include notes on how and when the formulae should be used. The students can then try to do exercises using their summaries as a source of information rather than their more bulky notes. If they find their summaries inadequate they can add to them. When they are revising for an examination they will be so familiar with some formulae that they can remove them, reducing the length of the list or using the extra space for more detail on topics they are less sure of.

Alternatively you could use a lecture or tutorial to revise a sample topic, helping students to construct an appropriate formula list as you proceed. Remind students that it is important to have well set out lists which are easy to read. Otherwise they will waste valuable time searching for hard-to-find information. This is one of the problems with open-book examinations: too much time is wasted turning pages.

By collecting the formula lists along with the examination scripts you can investigate one aspect of your students methods of learning mathematics. You will see which topics they have given most attention to and whether or not they actually made good use of their lists.

Reports 48

More and more importance is being placed on the ability to present the results of investigations in report form. For mathematics students report writing is required most frequently in final year projects, but in statistics reports can be used in earlier years to describe data.

It is quite probable that your students have had no previous experience with report writing. If they are specialising in mathematics or science they will be accustomed to answering questions mainly using symbols and calculations. They may have great difficulty in organising their ideas and writing coherent sentences. Even if your students have studied a communication subject as part of their degree they may have problems applying the recommended techniques in a new situation.

There is very little literature on report writing which is specific to statistics and possibly none for mathematics in general. Nevertheless there are a number of activities you could try to improve your students' report writing.

a   Give the students a lesson or a handout on report writing. A sample handout, which could be used as the basis of a lesson or given to students as a guide, appears as an appendix in *The Teaching of Practical Statistics*. This appendix provides further references on this subject.

b   Give the students some examples of reports or published papers which they can use as a basis for developing their own report-writing style.

c   Give students a report produced by a student in an earlier year. Ask students to criticise the report or even to mark it. Then ask them to rewrite it, making any necessary improvements.

d   Don't do anything initially, but each time students hand in reports use them to demonstrate the good and bad aspects of reports that they exemplify. Pass the marked reports around the class so that all the students can read the comments you have written. Discuss the reports with the class.

This approach is only appropriate if the students are handing in reports regularly. If there is only one report at the end of the course you will have to use one of the other methods so that students know what is required in advance.

Reference

Anderson, C.W. &   *The Teaching of Practical Statistics*
Loynes, R. M.   Wiley, 1987.

Assignments   49

Assignments are not usually a substantial component of assessment in mathematics courses. The reason is obvious. It is very easy for one student to produce a correct solution and for many students to copy it. How can assignments be used fairly and constructively?

Assignments which do not contribute to assessment

In this case there is little point in just copying someone else's solution. Also, if the assignments are marked in such a way that students can learn from their mistakes, that should be an incentive to hand work in.

Assignments which contribute to assessment

Here a delicate balance needs to be achieved. There need to be enough marks to make it worthwhile for students to hand in their assignments. There mustn't be too many marks or copying will give some students a large unfair advantage. It is also worth noting that even when a student copies a solution he or she may learn something about the problem in the process. By making assignments mandatory some learning will take place.

Computer generated assignments

When assignment questions are based on numerical information it is fairly simple to produce individual problems for a large number of students. This applies particularly in statistics and numerical analysis. The computer can generate data sets, matrices, etc. for each student as well as corresponding solutions for tutors. This simplifies part of the marking although these types of assignments generally involve some interpretation of the numerical results, comments on the methods applied and so on. Because each data set is different, the interpretation of results should vary from student to student.

Group assignments

These are usually only set in courses where the grade is pass or fail. It is assumed that if a student contributes to a satisfactory assignment she passes. Students in mathematics classes are often unused to co-operative activities, so some group work in tutorials may be needed to get the groups to function for assignment work.

Self-marked tests 50

How many times have you returned marked work to students, watched them glance at the grade and put the papers away? All the effort you put into making corrections and explaining misunderstandings has been ignored. Once markers observe this phenomenon a few times they loose their enthusiasm for constructive marking. When that happens there is not much for students to look at in their marked work. Part of the problem is that students' work is usually returned a week or more after it was handed in and by this time students may have forgotten any difficulties they encountered.

So there are two problems:

a   how to return work quickly so that the ideas are still fresh in the students' minds;

b   how to ensure that the student actually takes note of corrections.

Computer-marked tests satisfy (a) very well but do not provide useful comments at all. Some self-paced schemes provide tutors who mark tests as they are completed by individual students in the student's presence. This overcomes both problems but requires a large number of tutor hours.

A third possibility is to ask students to mark their own tests or assignments as you work through them in class. This may be difficult in a large group as students may be tempted to cheat. In a group of up to 25 or so this is less of a problem as it is possible to treat the students as individuals.

At the beginning of the course students can be encouraged to make sensible corrections by rewarding particularly good examples with a token grade or by public acknowledgement. Also, if you explain to the students that the purpose of making good corrections is to help them use their tests for revision they will quickly develop the habit of correcting their work well. Frequently when a student sees the correct solution together with her error she will gain a fresh insight on some point. If that insight is recorded immediately as a correction it will be fixed (at least temporarily) in the student's mind. For example, a student finding the cube roots of -8i made a slip and wrote the roots as

$$Z_0 = 2 \exp\left(\frac{-\pi i}{6}\right), \quad Z_1 = 2 \exp\left(\frac{\pi i}{2}\right), \quad Z_2 = 2 \exp\left(\frac{5\pi i}{6}\right)$$

and drew the Argand diagram.

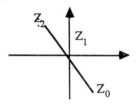

Alongside the diagram he wrote the correction:

Should have realised it wasn't symmetrical.

Self-marking has many advantages. It can provide immediate feedback, not just right or wrong, and focuses the students' attention on the kinds of errors they are making, so that they cannot ignore their own written comments. It also gives students some insight into how grades are awarded in examinations - very useful information.

It is a good idea to ask students to mark in a different coloured pen so that the original attempt and the corrections are easily distinguishable. Collecting the marked scripts for awarding grades or just for checking is also useful and doesn't take up much time. It also encourages students to make good corrections, discourages cheating and demonstrates that you are interested in seeing their work.

Uniform marking   51

Mathematicians are quite obsessive in their desire for completely objective assessment. This severely restricts the kinds of question that may be used for assessment and explains lecturers' reluctance to assess group work, assignments, oral presentations and the like. Even in tests composed of standard questions this complete objectivity is not always achieved.

If you have a number of tutors regularly marking work for their tutorial groups the amount of feedback they provide to the students can vary considerably, even if the marks they award are fairly consistent. Also, your tutors may be graduate students with little experience in marking constructively. You probably provide a detailed marking scheme when others are marking for you, but this may not tell the whole story. Students can also be encouraged to question their marks in a constructive manner so that they can find out exactly what was required of them, as well as checking the fairness of the marking.

If you plan to hold a workshop for your tutors at the beginning of the course, as described in Item 25: *Training tutors*, you could incorporate a session on assessment. A useful device for initiating discussion of assessment is to give the tutors copies of a number of pieces of students' work to assess. Ask the tutors to mark and annotate the work and then the whole group can compare the marks and comments. Or you could give out work that has already been marked, perhaps badly, and ask for comments and improvements. Either way, some consensus can be reached on the approach to be taken for the whole course.

A second session once the course is under way may be useful to deal with previously unforeseen problems.

Open-book examinations 52

This kind of examination is not widely used in mathematics for reasons that are not difficult to find. Mathematics examination papers tend to be rather stereotyped and frequently contain questions which are very similar or even identical to questions that have been used as examples in lectures or to questions that have appeared as tutorial exercises. In open-book examinations students are usually allowed to bring any materials they wish into the examination room and this includes solutions to the questions just mentioned.

Also, mathematics examinations often contain theorems to be proved and these are invariably to be found in texts or lecture notes. Finally, mathematics texts contain answers and often part solutions to exercises.

All these factors make it difficult though by no means impossible to set open-book mathematics examinations. Open book examinations are most suitable in courses involving problem solving or mathematical modelling where it is understood that the questions will be unrehearsed. However, these courses often don't involve any examinations and are assessed by means of classwork and assignments.

As an alternative to a complete open-book examination you could use the approach in Item 47: *Make up your own formula list* but extend the idea so that students can bring an extended course summary into the examination. You will need to specify what may or may not form part of the summary and how large it should be.

If you decide to try one of these approaches don't forget to give your students plenty of warning or you will risk creating an examination disaster.

Past examination papers often become the *de facto* syllabus for a course. Some students prefer to study past papers rather than working through the exercises you set during the year. The more stereotyped the questions are, the easier it is for students to pass an examination by restricting their study to past examination questions.

You will need to tell your students right at the beginning of the course how you intend to examine it. Then at suitable times during the course you should remind them about the form of the examination. And, towards the end, you should give them a sample paper or at least an outline of one. You also need to explain to the students why you think the examination needs to be changed. Try to convince them that your examination will reflect what you think is important about the subject and how it should be learned, and that the changes are in their interest.

Objective tests 53

Objective tests have become very popular both in a classroom situation where students mark cards and in computer laboratories where they type their responses directly. The items are most frequently of the multiple-choice type but can also include matching and true/false types.

Some advantages of objective tests are:

a   Large groups of students can be tested frequently with a minimum of effort

b   The record keeping can be done automatically along with the marking

c   The marking is consistent, there is no marker bias

d   Constructing the test items forces you to focus on what it is that you are really trying to test

e   If the test is taken at a computer the student obtains the result immediately. Even if the test is answered on cards, the answers or solutions can be given as soon as the test is finished

Some disadvantages are:

a   Great care should be taken in the setting and pilot-testing of objective test items, otherwise students may be able to deduce the correct response using a logical process of elimination and not much knowledge of the subject matter. For example in this question:

A sample of size 100 is chosen from a population with unknown distribution and with mean 30 and s.d. 16. The distribution of the sample mean will be:

147

> *a*   normal with mean 30 and s.d. 16,
> *b*   normal with mean 100 and s.d. 4,
> *c*   approximately normal with mean 30 and s.d. 1.6,
> *d*   approximately normal with mean 100 and s.d. 4.

A student who reads carefully will choose 30 as the mean and may decide that option (c) makes use of the given 100, avoiding the decision about normal/approximately normal altogether.

b   Students do not know where they went wrong. This is especially the case with numerical problems or problems that contain a large number of steps. The best way to overcome this problem is to avoid setting questions of these kinds and to use questions which test understanding of concepts or questions which show the steps in an argument in which the students have to find the errors.

An example of an item which tests understanding of a concept is:

> *It is true to say that any homogeneous system of linear equations $Ax =$*
>
> *a*   always has a unique solution;
> *b*   always has a solution;
> *c*   never has a solution;
> *d*   always has an infinite set of solutions.

Examples in which students have to locate errors tend to be lengthy, so the won't be included here. It is important to make sure that the errors involv major steps, not simple algebra, if the purpose of the question is to te something other than simple algebra.

c   There is a temptation to test only those aspects of the subject that ler themselves to objective testing and to ignore other aspects for which may be difficult to find suitable distractors. When you use distracto they should be genuine misconceptions. If that is not possible use a different type of objective question which does not require distractors, e.g. matching. For further information on writing objective questions see: *53 Interesting Ways to Assess your Students*.

d   Students do not get any practice in the written communication of their work, so it is essential to provide some other form of assessment in addition to objective tests.

e   Students can become very frustrated if they continually give wrong answers but don't know why. To overcome this you could provide solutions to tests and opportunities for students to discuss their errors with each other and with their tutors.

Some special difficulties arise when computerising objective tests, for example,

f   Although it is now possible to present questions on the screen in standard mathematical form this facility is not available to the student. If students are required to respond in symbols they have to use computer notation such as:

$$(x * y) \quad \wedge \quad 2.$$

g   When students are asked to provide an answer rather than choosing one is that there are usually many correct equivalent forms of an expression. It is possible to allow for some of these alternatives in the program but there will often be others that are not provided for. This problem is not quite so serious when the student is choosing from given answers, as in this case he has the opportunity to try to transform his answer to match one of those given.

Some of these computer-related problems will no doubt be solved in time. However, it is important to keep your educational objectives in mind when deciding to use objective tests so that the tests address the objectives.

Reference

Gibbs, G.,   *53 Interesting Ways to Assess your Students*
Habeshaw, S. & Habeshaw, T.   TES 1986.

Books available from TES

53 Interesting Things To Do In Your Lectures
Graham Gibbs, Sue Habeshaw, Trevor Habeshaw

53 Interesting Things To Do In Your Seminars And Tutorials
Sue Habeshaw, Trevor Habeshaw, Graham Gibbs

53 Interesting Ways To Assess Your Students
Graham Gibbs, Sue Habeshaw, Trevor Habeshaw

53 Interesting Ways Of Helping Your Students To Study
Trevor Habeshaw, Graham Gibbs, Sue Habeshaw

53 Interesting Communication Exercises For Science Students
Sue Habeshaw, Di Steeds

53 Interesting Ways To Appraise Your Teaching
Graham Gibbs, Sue Habeshaw, Trevor Habeshaw

53 Interesting Ways To Promote Equal Opportunities in Education
Vicky Lewis and Sue Habeshaw

53 Interesting Ways To Write Open Learning Material
Phil Race

253 Ideas For Your Teaching
Graham Gibbs, Trevor Habeshaw

Interesting Ways To Teach: 7 'Do-it-yourself' Training Exercises
Trevor Habeshaw, Graham Gibbs, Sue Habeshaw

Creating A Teaching Profile
Graham Gibbs

Getting The Most From Your Data
Judith Riley

TES books can be ordered from:
Plymbridge Distributors Ltd., Estover Road, Plymouth, PL6 7PZ.
Tel. 0752 705251   Fax. 0752 777 603